YOUR KNOWLEDGE HAS VALUE

- We will publish your bachelor's and master's thesis, essays and papers

- Your own eBook and book -
sold worldwide in all relevant shops

- Earn money with each sale

Upload your text at www.GRIN.com
and publish for free

The Fundamentals of Inorganic Polymers

Biswash Khatiwada

Bibliographic information published by the German National Library:

The German National Library lists this publication in the National Bibliography; detailed bibliographic data are available on the Internet at http://dnb.dnb.de.

ISBN: 9783346989888
This book is also available as an ebook.

Tribhuvan University

Institute of Engineering

Pulchowk Campus

Inorganic Polymer

Submitted By:

Name: Biswash Khatiwada

Submission Date: 2021-06-30

Abstract:

The first part of this paper focuses on the sources and way of manufacturing inorganic polymers. All the information has been gathered from book, articles and internet websites. Second part of term paper covers in details about daily used polymers and their applications and this paper also explains the importance of biodegradable polymers.

Introduction:

The word polymer is derived from two Greek words "poly" and "meros" which means many units. Polymer is defined as a molecule having high molecular mass (10^3-10^7U).It is also referred as macromolecules. Polymers are giants molecules that are formed by the joining the many basic repeated structural units on a large scale and those repeated structural units are derived from basic structure or simple molecule is known as monomer. Monomer is the reactive molecules and after passed through a large series of controlled reactions, polymers are obtained. Basically these simple molecules are linked with each other or forms networks or connected to chains by covalent bonds. The reaction converting monomers into polymers are called polymerization[2] and the number of repeating units (n) in a polymer is called degree of a polymerization[4].

For example, polyethylene is a polymer formed by linking together large number of ethane molecules[1].

$$nCH_2CH_2 \longrightarrow (CH2CH2)_n$$

Ethene (monomer)   Polyethylene(polymer)

The polymer is mainly classified into two types on the basis of source from where they are derived [1]. They are:

i. Natural Polymer :

These polymers are found in nature i.e. in plants and animals. For example starch, cellulose, proteins, rubber, etc [1].

ii. Synthetic Polymer :

Synthetic polymers are manmade polymers and are synthesized chemically to imitate the natural polymer. For example, Plastic, Nylon (6, 6), etc [1].

The polymer molecule consists of a "skeleton" (which may be a linear or branched chain or a network structure) and peripheral atoms or atom groups [3]. Synthetic polymer is further sub divided into two types on the basis of the atom present on skeleton. They are inorganic polymers and organic polymers [1].

1. **Inorganic Polymers:**

Inorganic polymers are those polymers which are composed of macromolecules composed of atoms other than carbon in backbone chain or in skeleton. These atoms are linked together by covalent bonds [1].

2. **Organic Polymer:**

Organic polymers are those polymers constituting the carbon atoms in skeleton or in backbone chain. The majority of synthetic polymers are organic polymers [1].

Differences between Organic and Inorganic Polymers:

i. Organic polymers contain carbon atoms in their backbones whereas other does not include carbon atoms in their backbone.

ii. Organic polymers are flammable and often releases toxic flames while latter one is not inflammable expect sulphur.

iii. Organic polymers are eco-friendly as they are degradable where as inorganic are not.

iv. Inorganic polymers are stronger (stiffer) but harder and more brittle than organic polymer.

v. Organic polymers have lower softening point as compared to inorganic polymers.

vi. Inorganic polymers contains many polar units and hence they are soluble in polar solvents while organic are not [1].

Body:

There are many varieties of inorganic polymers but among all frequently known polymers are:

1. Silicon Based Polymers

2. Phosphorus Based Polymers

3. Sulpher Based Polymers

1. **Silicon Based Polymers :**

It is also known as silicones or polysiloxanes. It is the organo-silicon polymers containing Si-O-Si linkage. It can be categorized as both organic as well as inorganic polymers.

These polymers resemble inorganic polymers in having high percentage of ionic character of silicone-oxygen bind and backbone chain contains atoms other than carbon. They resemble organic polymers in having groups on silicon atoms. Silicones may be cyclic silicones, chain silicones and cross-linked silicones [1].

Preparation of Silicones:

Preparation of Silicones involves following three steps and steps in details are shown below.

$$Chlorosilocones \rightarrow Silicols \rightarrow Silicones$$

i. **Preparation of alkyl or aryl substituted silicon chloride (Preparation of chlorosilanes)**

By the action of Grignard reagent with Silicon Chloride we can get a mixture of chlorosilanes.

$$1RMgCl + SiCl_4 \rightarrow RSiCl_3 + MgCl_2$$

$$2RMgCl + SiCl_4 \rightarrow R_2SiCl_2 + MgCl_2$$

$$3RMgCl + SiCl_4 \rightarrow R_3SiCl + MgCl_2$$

Mixture can be separated by careful fractional distillation since they have different boiling points [1].

ii. **Hydrolysis of alkyl or aryl substituted silicon chloride (Preparation of silicols or silanols or silandiols)**

It is prepared by heating alkyl or aryl substituted silicon chloride with water. These intermediates are very reactive and inflammable; they are preserved under dry inert gas such as nitrogen [1].

$$RSiCl_3 + H_2O \rightarrow RSi(OH)_3 + HCl$$

$$R_2SiCl_2 + H_2O \rightarrow R_2Si(OH)_2 + HCl$$

$$R_3SiCl + H_2O \rightarrow R_3Si(OH) + HCl$$

The hydrolysis of chlorosilanes is done to form the siloxane (silicone) backbone. Hydrolysis produces short-chain polymers which must then be lengthened to give the required molecular weight. Pure dimethyl dichlorosilanes are hydrolyzed to obtain oils [5].

iii. **Polymerization of silicols:**

Silicones are obtained by condensed polymerization of Silicols. Type of polymerization depends upon the nature of R and the way in which hydroxyl group undergo polymerization. Long straight chain, cyclic chain and crossed-linked polymers can be obtained from condensed polymerization [1]. Structure of linear polymer is shown in figure:

$$nHO-\underset{\underset{CH_3}{|}}{\overset{\overset{CH_3}{|}}{Si}}-OH + HO-\underset{\underset{CH_3}{|}}{\overset{\overset{CH_3}{|}}{Si}}-CH_3 \xrightarrow[-H_2O]{Polymersatia} -O\left(-\underset{\underset{CH_3}{|}}{\overset{\overset{CH_3}{|}}{Si}}-O\right)_n-\underset{\underset{CH_3}{|}}{\overset{\overset{CH_3}{|}}{Si}}-CH_3$$

Silicone

The addition between silanols enables the polymer chain ends to be de-activated i.e. to give non-reactive oils [5].

Properties of Silicones:

i. Depending on the proportion of various alkyl silicon halides used during the preparation, silicones may be liquid, viscous liquids, semi-solid (grease), rubber and solids.

ii. Because of the oxygen-silicon bonds (bond dissociation energy = $502 kJmol^{-1}$), they exhibit outstanding stability at high temperature.

iii. Because of the presence of alkyl groups surrounding the silicon atoms makes the molecule hydrophobic and gives water repelling tendency of silicones.

iv. Their physical properties are less affected by the change of temperatures.

v. They are non-toxic [6].

Uses of Silicones:

i. Straight chain polymers up to 500 units are silicones fluids. They are used as water repellent for treating buildings, glassware and fabrics. They are also included in car and shoe polish.

ii. Silicones fluids are non toxic and have low surface tension; therefore, addition of a few parts per million of it greatly reduces foaming in sewage disposal, textile dyeing, beer

Page | 5

making and frothing of cooking oil in making potato crisps and chips. Silicone oils are used as dielectric insulating material in high voltage transformers. They are also used in hydraulics fluids.

iii. Silicone greases are made by mixing silicones oil with lithium stearate soaps and are used as lubricants where high and low temperatures are encountered.

iv. Silicones rubbers are obtained by mixing long chain silicones with filler like silicon dioxide in the presence of curing agent like peroxide. Rubber has 6000 to 600000 Si units. Silicon rubber is useful because they retain their elasticity from -90.C to +250.C, which is much wider range as compared to natural rubber. They are used in making tyres of aircraft and racing cars, as sealing material in search lights and aircraft engines, in making lubricants, paints and protective coatings. For making boots for the use at very low temperature (Neil Armstrong used silicones rubber boots while walking on the moon).

v. Silicones resins are used in making high voltage insulator, making printed circuit boards and to encapsulate integrated circuits chips and resistors.

vi. They are also used as nonstick coatings for pans, moulds for making car tyres etc [6].

2. Polyphosphazines

Nitrogen and Phosphorus have little tendency to their own to undergo catenation. However nitrogen and phosphorus catenate together resulting very extensive chemistry of polymeric compounds known as polyphosphazines. Simply polyphosphazines are inorganic polymers containing phosphorus atoms [1]. Their general formula is

$$\left[-N = P\!\!\begin{array}{c} R \\ | \\ | \\ R \end{array} - \right]_n$$

Where,

    R = -Cl in polyphosphonitrillic chlorides

    R = -OCH$_3$ in polydimethoxy chlorides

    R = -OC$_2$H$_5$ in polydiethoxy chlorides [1].

Preparation of polyphosphonitrillic chlorides:

Polyphosphonitrillic can be prepared by the reaction between phosphorus pentachloride and ammonium chloride in presence of c_6H_5Cl or by heating PCl_5 with NH_4Cl at 120°-150°C [1].

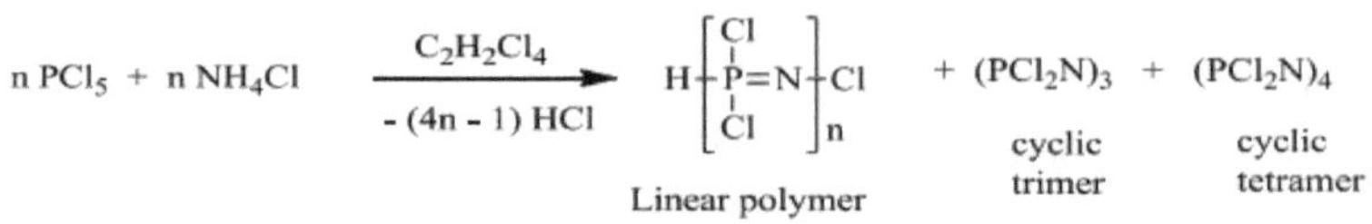

$$n\ PCl_5 + n\ NH_4Cl \xrightarrow[- (4n-1)\ HCl]{C_2H_2Cl_4} H\left[\begin{array}{c} Cl \\ | \\ P=N \\ | \\ Cl \end{array}\right]_n Cl + (PCl_2N)_3 + (PCl_2N)_4$$

Linear polymer   cyclic trimer   cyclic tetramer

The trimeric or tetrameric compounds are easily separated by distillation method [1].

Structure of cyclic polyphosphonitrillic chloride can be represented as:

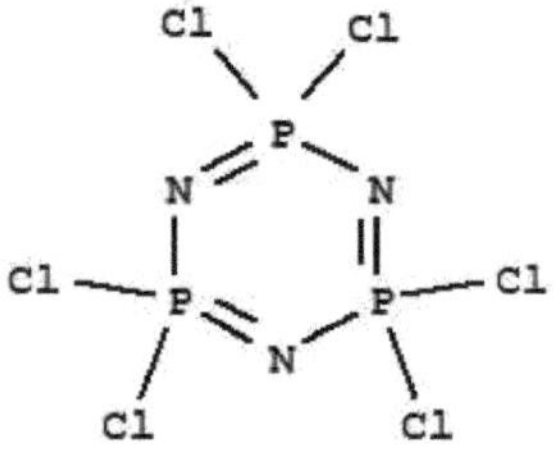

Properties:

i. These polymers exhibit high elasticity and can be stretched many times reversibly hence, also called inorganic polymers.

ii. Freshly prepared samples are soluble in chloroform, but are insoluble in hexane. However, when the solution is allowed standing, they slowly form gel due to cross-linking.

iii. Their storage in the absence of air does not cause any change in their elastic properties. However, in the presence of moisture, the polymer becomes brittle due to the formation of oxygen bridges between chains [6].

Polydimethoxy and Polydiethoxy phosphazines

Polydimethoxy and polydiethoxy phosphazines are prepared by treating polyphosphonitrilic chlorides with sodium methoxide and sodium ethoxide respectively [1].

Properties:

Page | 7

i. They are colorless, odorless, transparent and film-forming thermoplastics.

ii. On heating above 100.C, they slowly form cyclic polymers [1].

Uses:

Some of the important properties of polyphosphazines are listed below:

i. They are used as rigid plastic, expandable foam and fibers.

ii. Certain type polymers show water repellant properties and is used to make water proof thermoplasts [1].

3. Sulphur Based Polymers:

Sulpher because of its ability to from catenation forms open and cyclic S_n species. So, a large numbered of sulphur based polymers is known and are classified as following [1]:

A. Linear chain polymer

i. **Polymeric Sulphur (PS):**

It is prepared by heating rhombic sulphur at 165oC to 168oC following by quenching the molten mass on ice bath. The so called plastic sulphur is obtained [1].

Uses:

a) It is to make flame proofing fabrics i.e. as plasticizers.

b) It is also used as catalyst in the manufacture of silicons [1].

iii. **Polymeric Sulpher nitride $(SN)_n$:**

Polythiazyl also known as polymeric sulfur nitride $(SN)_n$ is an electrically conductive gold coloured polymer with metallic luster. It was first discovered around 1910 as the first conductive inorganic polymer. Examples are disulfur dinitride ($S_2 N_2$; a dimer), tetrasulfur tetranitride ($S_4 N_4$; a tetramer) etc [6].

Uses:

a) Due to its electrical conductivity, it is used to make digital circuit, computers, electric power transition, transformer, power storage devices, electric motor etc [1]..

B. Network Polymers:

Sulpher based network polymers are cross-linked polymers formed by reacting polyvalent elements such as silicone antimony, arsenic, bismuth, germanium, indium etc. with chalcogens(ore forming: sulphur, selenium and tellurium). The resulting binary compounds are of great variety and complexity is known as chalcogenide glasses (or chalconide glasses). The best known chalcogenide glass is $(As_2S_3)_n$ [1]. Chalcogenide glasses are prepared by melt-quenching or passing the chalcogens compounds vapour over the polyvalent elements like arsenic, antimony, indium, germanium, phosphorus, tin and thallium. [6].

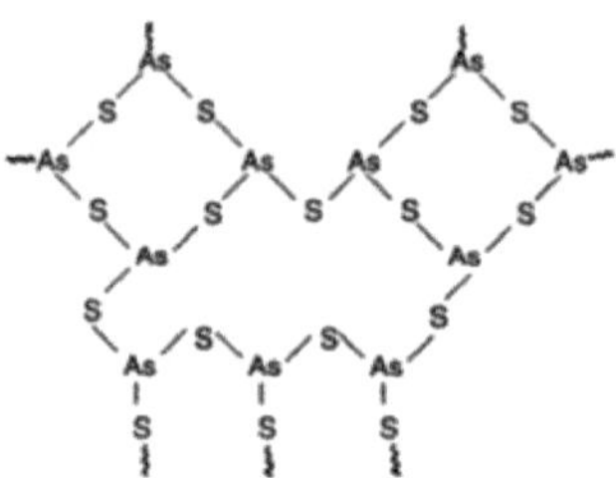

Figure of $(As_2S_3)n$

Properties:

i. They resemble organic polymers as well as glasses. However, they possess lower softening point, tensile strength and higher refractive indices than borosilicate glasses.

ii. They are fairly stable to acids, bur are attacked by concentrated alkalis.

iii. They oxidize in air at about 3000oC. Therefore, these polymers can be distilled in vacuum without any decomposition.

iv. They are deeply colored and most of them transmit infrared radiations.

v. Their conductivity changes reversibly from low to high value under applied voltage. This phenomenon is known as switching [6].

Uses:

i. They are used in ultrasonic delay lines, high energy particle detectors, mouldable infrared optics such as lenses, and infrared optical fibers.

ii. They are widely used in rewritable optical disks and phase-change memory devices and also used in fabrication of IR for army and civil optical devices [6].

Organic Polymers:

Page | 9

These polymers skeleton structure are mainly made up of carbon atoms. The atoms attached to the side valencies of the backbone carbon atoms are usually hydrogen, oxygen, nitrogen etc [1]. Organic polymers are classified on the following types on the basis [6]:

a. **Source:**

Depending on the source, organic polymers are classified into two major types [6]:

i. **Natural:**

Examples are: starch (polymer of alpha-D Glucose), cellulose (polymer of beta-D Glucose), proteins (polymer of amino acids), natural rubber (polymer of isoprene) etc [6].

ii. **Synthetic Polymers:**

Examples are: Polythene, Teflon, Bakelite, Kevlar, etc [6].

b. **Methods of synthesis:**

Depending on the method of synthesis, polymers are classified into two main types:

i. **Addition Polymers:**

In addition polymers all the atoms in the monomer are used to form the polymer. Generally alkenes or alkynes form addition polymers [6].The polymer product, formed without the loss of any small molecule, is exactly the multiple of original monomeric molecule. Application of energy in the form of heat, light, pressure or presence of catalyst [1]. Examples of addition polymers are polythene (polyethylene), polypropylene, polyvinyl chloride (PVC), polytetrafluoroethylene (Teflon), polystyrene, etc.

Polyethylene

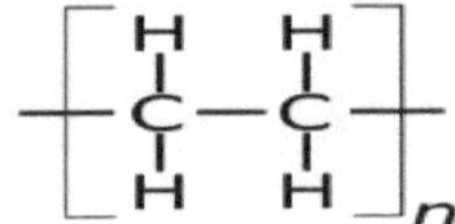

ii. **Condensation Polymers:**

It is defined as the reaction between functional molecules leading to the formation of a polymer [1]. In condensation polymers monomers join up with an expulsion of small molecules like water, ammonia, methanol, etc (all the original atoms are not present in the polymer) [6]. Polyamides and polyesters are two important classes of condensation polymers. For example condensation between phenol and formaldehyde formed Bakelite [1]. An example of condensation reaction is shown below:

$$n \; \underset{HO}{\overset{O}{\parallel}}C-R-\underset{OH}{\overset{O}{\parallel}}C \; + \; n \; H_2N-R'-NH_2 \longrightarrow \left[-C-R-C-N-R'-N- \right]_n + 2 \; H_2O$$

c. **Molecular Force:**

Depending on the intermolecular force, organic polymers are classified as:

i. **Thermosetting polymer:**

Thermosetting plastics are those which are set with heat and have little elasticity. Once set, they cannot be reheated and reformed. They are heated and molded during manufacture. Once cooled, they will not soften again when heated. This breaks the potentially unending cycle that thermoplastic plastics are capable of. If heated too much, they burn. Thermosetting plastics also have lots of long chain molecules, but there are links between them. These cross links prevent the molecules from moving over one another. Melamine, Bakelite, etc are the examples of thermosetting plastics [6].

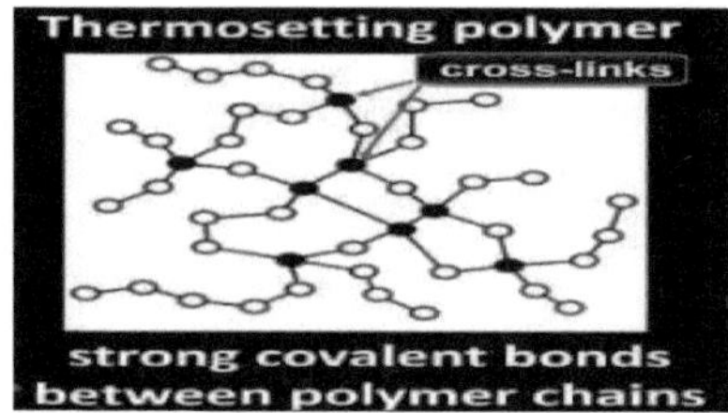

ii. **Thermoplastic polymer:**

The majority of common plastics are thermoplastics. Thermoplastics can be heated and reshaped because of the ways in which the molecules are joined together. Reheating and reshaping can be repeated many times (as long as no damage is caused by overheating). Thermoplastic plastics are made of long chains of polymers, which don't cross over very often. When heated, the molecules slip easily over one another. Polyethylene, polystyrene, polyvinyl chloride, etc., are the examples of thermoplastics [6].

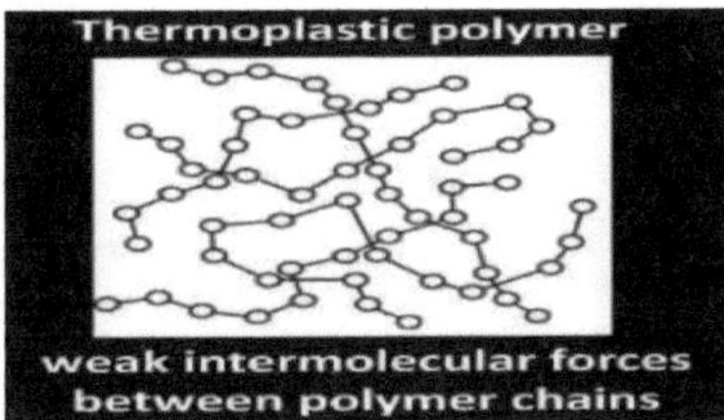

Some examples of additional polymers:

a. **Teflon:- Polytetrafluoroethylene, Fluon (PTFE)**

Polytetrafluoroethylene is obtained by polymerization of water-emulsion of tetraflouroethylene, under pressure in the presence of benzoyl peroxide [1].

$$n \; \underset{\underset{F}{|}}{\overset{\overset{F}{|}}{C}} = \underset{\underset{F}{|}}{\overset{\overset{F}{|}}{C}} \quad \xrightarrow{\text{benzoyl peroxide}} \quad \left[\underset{\underset{F}{|}}{\overset{\overset{F}{|}}{C}} - \underset{\underset{F}{|}}{\overset{\overset{F}{|}}{C}} \right]_n$$

tetrafluoroethylene   polytetrafluoroethylene
(TEFLON)

Uses:

i.   Teflon is mostly used as insulating materials for cables, wires, transformers, etc.

ii. They are used in making gaskets, stop cocks in burettes and pipes for carrying chemicals.

iii. They are also used as coating in high quality non-stick cooking pans.

iv. They are also used in the making of self lubricating bearings [6].

b. **Polyvinyl chloride (PVC)**

PVC is obtained by heating a water-emulsion of vinyl chloride in the presence of small amount of benzoyl peroxide in an autoclave under pressure [6].

Uses:

PVC is used for making sheets, light fittings, safety helmets, cycle and motorcycle mudguards, pipes and pipe-fittings, chemical container, frames for doors and windows, debit and credit cards, etc [6].

c. **Polystyrene (Polyvinyl benzene):**

Polystyrene is prepared by polymerization of styrene (dissolved in ethyl benzene) in presence of benzoyl peroxide as initiator (catalyst) [6].

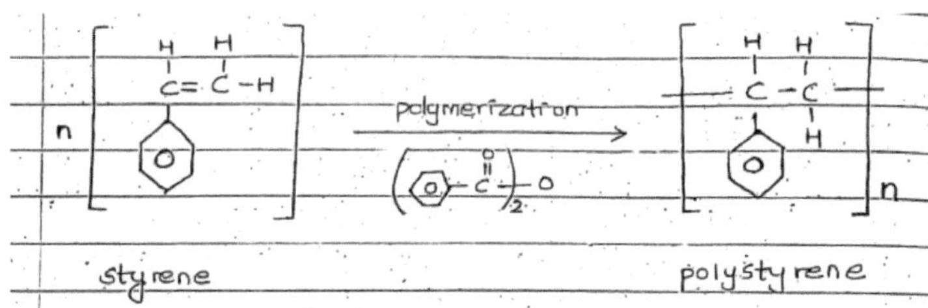

Uses:

i. Polystyrene is used in protective packaging and containers as styrofoam, toys, disposable food container.

ii. They are also used as insulators.

iii. They are used in making lenses and disposable cutlery like spoons, forks, plates, trays.

iv. They are used in making radio, TV and refrigerator parts, CD cases, etc [6].

Some example of condensation polymers:

a. **Bakelite:**

Bakelite is phenol-formaldehyde resin, a condensation polymer formed by polymerization of phenol with formaldehyde. Bakelite is prepared by condensing phenol with formaldehyde in the presence of acidic or alkaline catalyst [6].

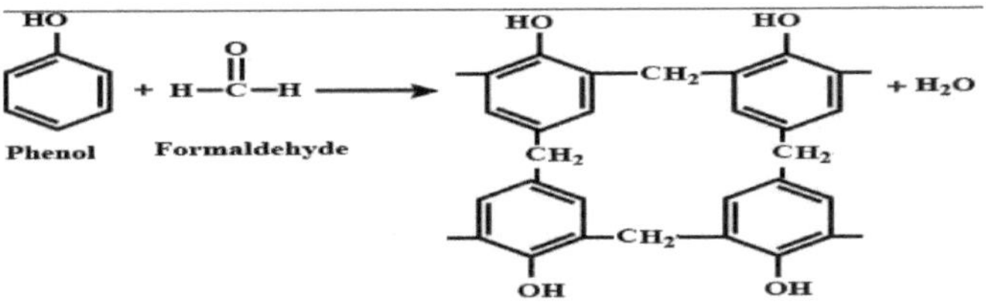

Uses:

i.   It is used for making electric insulator parts like switches, plugs, switch-boards, heater-handles, etc.

ii. It is also used for making telephone parts, cabinets for radio and television.

iii. It is also used as hydrogen-exchanger resins in water softening, etc [6].

b. **(Nylon-6,6):**

Nylon-6, 6 is a polyamide and is obtained by polymerization of adipic acid with hexamethylene diamine under nitrogen atmosphere at 200°C [6].

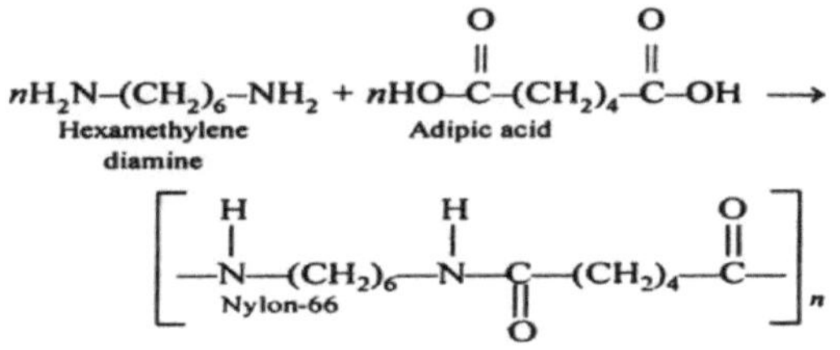

Uses:

i.   It is used for ball bearing cages, electro insulating elements, pipes, profiles and various machine parts.

ii. Other popular applications are: carpet-fibers, apparel, airbags, tyres, zip ties, opes, conveyor belts, under-garments, dresses, carpets, hoses and the outer layer of turnout blankets [6].

c. Polyurethane (Polyvinyl benzene):

Polyurethanes are obtained, commercially by treating diisocyanate and diol. For example when methylene diphenyl diisocyanate (diphenylmethane diisocyanate; MDI) is treated with ethane-1, 2-diol we get polyurethane polymers [6].

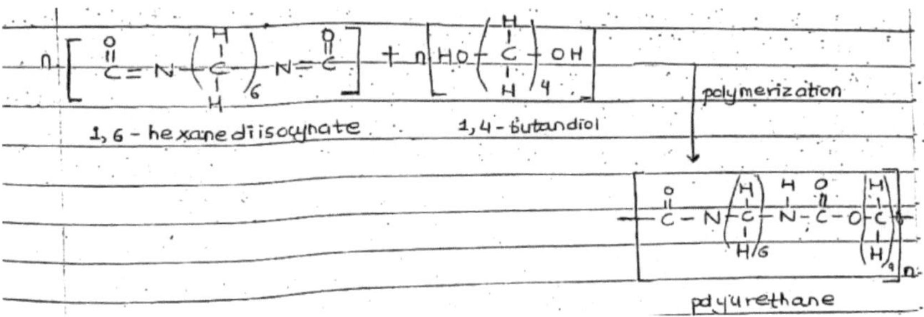

Uses:

i. Polyurethanes are used as coatings, films, foams, adhesives and elastomers.

ii. Resilient polyurethane fibres (spandex) are used for foundation garments and swim-suits.

iii. They also find use as a leather substitute as corfoam.

iv. They are used in cast to produce gaskets and seals [6].

d. Epoxy Resins:

Epoxy resins are materials prepared by a polymerization in which one monomer contains epoxy groups [6].

Uses:

i. As surface coatings, adhesives, for skid resistant surfaces for highways, moulds made from epoxy resins are employed for the production of components for aircrafts and automobiles. [6].

The problem of plastic waste arise as harmful material responsible for environment pollution. Products where the plastics are used for a short time before becoming waste are difficult to recover after use and remain in the environment [7]. Because plastic are shielded from sunlight in landfills, they will not decompose for thousands of years. Plastic needs about 450 years just to start decomposing. Then, it takes another 50-80 years to decompose completely [6]. So, to reduce the amount of plastic or non-degradable polymer accumulated in environment i.e. to reduce the environmental pollution and its harmful effect on ecosystem manufacture of biodegradable polymer was necessary.

Biodegradable Polymers:

The polymers which are degraded by micro-organisms such as bacteria, fungi and algae within a suitable period so that the polymers and their degraded products do not cause any serious effects on the environment are called biodegradable polymers. For example, when ethylene is polymerized in the presence of carbon monoxide, a polyethylene like polymer that contains carbonyl group is produced. This material, which has been used in the production of trash bag undergo degradation upon exposure to UV due to carbonyl group.Some examples of biodegradable polymers are polylactic acid, polyvinyl alcohol , starch derivative, cellulose etc [6].

Application:

i. Biodegradable polymers are used mainly for medical goods such as: Surgical sutures, tissue in growth materials or for controlled drug release devices, plasma substitutes etc.
ii. These are also finding use in agriculture materials (such as films, seed coatings), fast food wrappers, personal hygiene products etc [6].

Non-biodegradable polymer:

A polymer whose degradation is not possible by the action n of biological enzymes or naturally occurring microorganism such as bacteria, fungi and algae is called non-biodegradable polymer. Such type of polymer is not decomposed by, micro-organism or biological factor like oxygen, heat radiation, ozone, moisture etc [1].

Most of the synthetic polymers are non-degradable polymers. Xylon, nylon, polyester, etc are synthetic as well as non-degradable polymers [1].

Conducting Polymer:

Most polymeric materials, due to lack of free electrons, are insulators. Within the past few years, polymeric materials have been synthesized which posses electrical conductivity. A polymer which can conduct electricity is known as conducting polymers [1].

Polyacetylene has conjugated double bonds but is not a conductor; however, by a process called doping, which involves introducing small amounts of electron-donating or electron accepting compounds, it is possible to produce polyacetylene to show conductivity. The purpose of the doping agent is either to remove electrons from the pi system (p-doping) or add electrons to the pi system (n-doping). In the above example polyacetylene is p-doped [6].

Applications of conducting polymer:

i. In rechargeable light weight battery.

ii. In wiring in aircrafts and aerospace company.

iii. In solar cells, electronic display and optical filters.

iv. In drug delivery system for human body [1].

Conclusion:

Polymers are everywhere in plastic, antibacterial plastics, where growth of bacteria is ideal , care products , food like starches, proteins gum, manufacture of automobiles, tyres, electrical insulating materials part and machines part and it has completely revolutionized the daily life as well as industrial sector. It is also even in our DNA. Since, it is a part of everyday lives and also have a wide range of application, therefore we must know about its sources or way of manufacturing polymers. Inorganic polymers attached with us in day to day life but majority of the inorganic polymers must be replaced by organic polymers to reduce the effects on environment.

Acknowledgement:

I would like to express my gratitude to my respected professor Dr. Ram Kumar Sharma for providing me an opportunity to write my first term paper. Thank you professor for motivating me.

Reference:

[1].Sharma and Panthi (2016), *A Textbook of Engineering Chemistry,* Heritage Publication.

[2].*NCERT* (Author and Publisher), 2019

[3].Van krevelen and Nijenhuis (2009), *Properties of Polymers.*

[4].Jain & Jain, *Engineering Chemistry,* Dhanpat Rai & Sons Publishing Company, 16th edition.

[5].*https://www.elkem.com/silicones/technologies/basic-silicones-products/chlorosilanes/*

[6].Prasai, Keshar (2017), *Polymer-Note.*

[7].*https://onlinelibrary.wiley.com/doi/10.1002/macp.202000017*

YOUR KNOWLEDGE HAS VALUE

- We will publish your bachelor's and
  master's thesis, essays and papers

- Your own eBook and book -
  sold worldwide in all relevant shops

- Earn money with each sale

Upload your text at www.GRIN.com
and publish for free